THE INTERNATIONAL TRACT

International Harvester came into being in 1902 as a result of the merger of a number of implement firms including McCormick & Deering. Until recently this has always been reflected in the fact that various combinations of the names McCormick, International and McCormick-Deering could be found on their tractors.

International produced its first tractors in 1905 but did not enter the field on a serious basis until 1910-11 with its Titan and Mogul ranges, these being basically developments of the steam tractors then used in the USA using large slow speed stationary engine type power units. The introduction of the 8-12 Mogul and 10-20 Titan models in 1914 heralded the move into smaller more manageable units. These were the last of the primitive models with horizontal hit and miss governed engines, chain steering and tank cooling.

As tractor production was only a part of the International activity, a UK office was established in 1909 and the need for tractors during the First World War meant that Titan and Mogul tractors were imported and sold into government service. International had also developed a range of "Hamilton" self lift ploughs for use with these tractors and such an advanced design for the time was to receive many imitators in years to come.

8/12 Junior

Reliable a unit as the Titan 10-20 was, there was no denying once the British (and for that matter the US) farmer had had the chance to try a Fordson, the Titan looked cumbersome, heavy and outdated. At 2.5 tons it was a heavy brute but IHC persevered and prices were reduced from the original £400 to £390 even throwing in a free plough into the bargain.

The shape of things to come however was to be seen in 1917 when the "Junior" 8-16 model was introduced. This weighed only 1.5 tons and although the general layout could not compete with the Fordson certain features incorporated were to have important effects later on.

As far as the transmission was concerned, the chain drive to the rear wheels was still evident, but three forward and one reverse speeds were provided. The engine was the most important new feature. It was a 4-cylinder unit with 4 inch bore and 5 inch stroke, but had overhead valve configuration plus removable cylinder sleeves. Ignition was by high tension magneto, and the most unusual feature was, of course, the mounting of the radiator behind the engine block in front of the driver, to avoid damage it is said!

10/20 and 15/30

The one piece frame method of construction was introduced in 1921 when the previous 15-30 model was ousted by a tractor with a vertical 4-cylinder engine sporting a ball bearing crankshaft, and for the first time, a power take-off as standard. A smaller version of this ousted the 10-20 Titan in 1922 and this 10-20 was to be the best known of all "International" tractors. It enjoyed a production run of seventeen years, the last being sold in the UK in 1942 although production had officially ceased in 1939. The tractor was also produced in industrial form.

The International 10-20 and 15-30 models were advertised as "Triple Power" tractors, featuring power at the drawbar, PTO and pulley. A compromise in unit construction between the "Wallis" and Fordson principles was adopted, the whole power train being housed in a cast iron hull, but the engine was a separate unit mounted therein. These tractors featured three forward gears of 2, 3 and 4mph, although a high road gear could be supplied to special order.

In 1929 major alterations were effected on the 15-30 model, which subsequently was known sometimes as, the 15-30, and others as the 22-36. These involved engine design changes and an improved drawbar.

The 15-30 was ousted by the W30 in 1934, whilst the 10-20 did not go out of prod
10-20 sported the red col
were not delivered to UK users until well
of course, the design had been out of production for nearly two years, but any farmer struggling in war-time conditions was glad to see as reliable a machine as the 10-20.

The 10-20 engine was simply a smaller version of the 15-30 and on test at the 1930 trials gave 23.4hp at the belt and 17 at the drawbar. With a bore and stroke of 4.5 X 5 inches it ran at 1,000rpm, the same as the original 15-30, the 22-36 being run at 1,050rpm. A single plate clutch was provided to take up the drive to the gearbox. Only the 15-30 model featured impeller assistance to the cooling system however, and the connections to the water pump can be clearly seen in the illustrations.

The original 15-30 engine was of 4.5 inch bore by 6 inch stroke and at the 1930 World Trials the late, larger engine of 4.75 inch bore was tested and gave a maximum of 39bhp at the belt and 29 at the drawbar. An unusual feature however, was the use of a two bearing crankshaft with ball bearings on the mains, the crank itself being of very heavy construction. An Ensign carburettor was fitted, along with a Dixie 46-C magneto on early models. This later gave way to International's own E4A unit. Replaceable cylinder liners were a feature inherited from the 8-16 Junior. Oil filtration equipment was not evident on some early models, and the lubrication system was a combination of splash and force feed.

The Farmall

The work of International designer Bert R. Benjamin first saw the light of day in 1923, at a time, incidentally, when International were still experimenting with steam! One of the problems encountered in tractor design was the difficulty in adapting the tractor to cultivate row crops. Much experimentation was carried out and numerous prototypes constructed, some as early as 1914. The result was the "Farmall" of 1923.

The tractor attained high axle clearance by virtue of it having the final geared reduction to the rear wheels housed in casings at the outer ends of the axle shafts.

The hull of the 10-20 was abandoned in favour of frame construction. Independant brakes were provided and when certain front wheel equipment was fitted these could be automatically applied. The front axle was mounted on a pillar "out front" of the radiator and the steering column ran along the top of the bonnet to meet it. A normal four wheel configuration could be provided or single or vee twins substituted.

The Farmall range were the first true "rowcrop" tractors and were to be copied to a greater (or lesser) extent by most other manufacturers. A complete range of implements were offered for use with the tractors, and a power lift attachment for the F20 would cost you £16. Most of the implements were designed to fit onto the tractor frame.

In developing the Farmall some of the early prototypes looked like motorised toolbars rather than tractors but once the design had become evolved, a tractor using the same basic units as the regular series was evolved, but with a channel frame rather than the cast hull of the 10-20 and 15-30 models. Independant brakes were provided, and these could be applied either by lever or automatically when the front wheels reached a certain lock.

Initial "Farmall" sales were slow, but by 1929 the model had become established, sales in the UK being encouraging also. The 3.75 X 5 inch bore and stroke of the engine, running at 1,200rpm gave 12hp at the drawbar. Three forward speeds were provided.

The success of the F20 led to the introduction of a slightly larger version, the F30 in 1931 - the designation F20 being bestowed upon the original Farmall at this time. The 15-30 was also made available in uprated form as the 22-36 and so continued until 1934.

In 1932 following the introduction of the F30 the model was given the F20 name and this featured a slightly larger build and had 4-forward speeds. The F20 cost £260 on steels in 1936 and was replaced by the Farmall H in 1939. The F30 on the other hand used a 4.75 X 5 inch bore and stroke and at 1,150rpm gave 20hp at the drawbar and 30hp at the belt, and would cost you £350 in 1936.

The smallest of the "Farmall" range appeared in 1932 - the F12 and this was followed by the W12 "Regular" model in 1934 using many of the F12s components including the engine. "Fairway" (or Golf Course) and "Orchard" versions of the F12 designated Fairway-12 and 0-12 respectively were also produced.

Using a 3 inch bore and 4 inch stroke, its 4-cylinder engine ran at 1,400rpm and gave 9hp rating at the drawbar and 14.6 at the belt. The speed gearbox featured the shafts mounted parallel to the rear axles and the rear wheel track was adjustable from 44 to 60 inches on splined shafts. The early examples featured a slightly different engine with the magneto on the left side of the tractor.

In 1936 the F12 tractor sold in the UK for £185 on steel wheels, £225 on rubbers, and with a single front wheel £178.

A wide front axle was available instead of the twin vee front wheels and during the production run the tractor became more popular on pneumatics, the rear tyre size being upped from 36 to 40 inches in 1937. The light French & Hecht spoked type wheels were the most suitable for this model, but Dunlop equipment was used in the UK.

The 1938-39 version of the F12 was the F14 which gave more power due to an increase in engine speed to 1,650rpm. Horsepower went up to 14.8 at the drawbar and 17.4 at the belt. The driving position was modified to give better visibility thus requiring a sloping steering column. The engine of the F14 also featured a fuel lift pump, International Harvester's own F-4 magneto, and A10 carburettor.

Although the F14 enjoyed a successful two years production, when a good number were also sold in the UK out of over 27,000 made it, and other Farmall models, were phased out with the new range coming out in 1939.

The W30

The W30, introduced in 1932, featured an engine similar to the 22-36 but with 4.25 inch bore and 5 inch stroke which, running at 1,150rpm, gave 31.3hp at the belt and 19.7 at the drawbar. Four forward speeds were provided however. The power unit was the same as that fitted to the W30 Farmall. W30 tractors were produced from 1932 to 1940, with the option of a 10mph top gear on pneumatics. International models were all available from late 1933 on rubbers, although the American style French & Hecht spoked centres were not common in the UK, most tractors receiving Dunlop or Firestone centres on importation, of the usual pattern seen also on Fordson, Marshall and other makes.

By 1932 International was using three plants for tractor production at Chicago. Milwaukee, and the old Moline Plow works at Rock Island. The latter being used for the production of Farmall models from 1927. In addition a whole range of implements were offered, and in the UK, in addition to offices at 259 City Road, London, there were workshops at Liverpool, a distribution depot in Leith, Edinburgh and an Irish office in Dublin. International Harvester in Britain was by then a British company wholly owned by the parent US concern but one which operated with a certain air of independence. All sales literature bore the London address and most was prepared in the UK.

TracTracTors

International commenced experiments with crawler machines in the mid-1920s and early examples were based on the 10-20 tractor. When the basic power units were available it was not a major task to build a crawler round them, and the 15-30 tractors were also modified. By 1930 a model 15 TracTracTor (as IHC crawlers were named) had been developed using a much improved steering system from the early models. The model 15 was not produced in quantity but in 1931 the T20 came out and this used the same basic engine as in Farmall F20 tractors. It had slightly more power as governed revs were 1,250.

Three forward speeds were provided, and the steering clutches and brakes were so mounted as to be readily accessible from the rear of the tractor.

Towards the end of 1932 two further models were added to the range. These used an new series of engine and were the T-40 crawler, and TD-40. Whilst the T-40 had an orthodox 6-cylinder unit, the TD-40 featured the first of a line of IHC diesels of special design. On looking at the engine, the right side sported carburettor, magneto and spark plugs. The left side however showed diesel injection pump and injectors. The secret lay in each cylinder having an auxiliary combustion chamber which was brought in to lower the compression ratio thus enabling starting on petrol, as each had a spark plug therein. Once speed was built up however the auxiliary chambers were cut out, fuel pump cut in, petrol shut off, and the tractor functioned as a normal diesel.

The T40 and TD40 crawlers used the same engines as the W40 and WD40 wheeled models respectively. Only 103 units came to Britain. The T40 could give 44hp at the drawbar although the rated horsepower of both the petrol and diesel models was quoted as 33. In the UK you could buy a TK-40 for £770 and a TD-40 for £995. Numerous accessories were available for the TracTracTor series. These included sump guards, a host of various track shoes and plates, wide or narrow tracks, road pads, direct dynamo lighting, or battery type lighting and starting. Cab and canopies were also available.

Two new crawlers also saw the light of day in 1936: the T35 and TD35 models. These being less powerful versions of the "40" series with reduced cylinder bore.

W Series - W40 and WD40

The largest wheel tractors to date, the W40 and WD40 came along in 1934.

The W40 was equipped with a 6-cylinder engine of 3.75 inch bore by 4.5 inch stroke, which running at 1,600rpm gave 48hp at the belt and 34.5 at the drawbar. With the higher engine speeds the ball bearing crankshaft was dispensed with, a 7-bearing unit using removable shells being fitted. In 1936 the WK40 which was the paraffin burning model would cost you £420.

The WD40 held the distinction of being the first diesel wheeled tractor to be offered in the USA, but in the UK there had been others before it arrived. The 4-cylinder engine of 4.75 inch bore by 6.5 inch stroke, while running at 1,100rpm gave 28hp at the drawbar and 44 at the belt. The fuel injection system was International's own whilst, like the W40, a three speed gear box was fitted. The tractor cost £590 in 1936 and turned the scales at over 3 tons.

New Colours

1936 saw a colour change for International tractors, these appearing in a red finish, having previously been grey with red wheels. The "Farmall" range were the last to change in November, but it was well into 1937 before red tractors were seen in the UK.

1938 saw the last of the old style models to appear with the updating of the F12 and W12 models to the F14 and W14.

All Change

1939 was "all change" for International and the first of a new range came out. The "Farmall" models were phased in first, and consisted of the "A" a one-plow rowcrop tractor with engine mounted left of centre line to permit clear ahead vision, and the similar "B" which was slightly more powerful but had a single front wheel. Some of these found their way into the UK just before the outbreak of war. The "H" and "M" models were the best known over here and numbers of these continued to be supplied under "Lease-lend" right through the war. The first of the new crawlers also appeared at this time, the TD18. It was 1940 by the time the rest of the new range appeared.

New Rowcrop Models

The 1939 introduction of the Farmall models, A, AV, and B ousted the old F-14 model. Both models A and B were to be seen in the UK. The engine of 3 inch bore and 4 inch stroke gave 17 rated bhp at the drawbar at 1,400rpm.

The Model A was equipped with a wide front axle, the power train being offset to the left side to enable a clear view ahead when working in row crops. A four speed gearbox was provided.

The Model AV was a vineyard model with high clearance, but only appeared in the UK in very small numbers. The Model B featured either a single or vee twin front wheel and both rear axle shafts were the same length, the driving position being still on the right side however.

The Models A and B were produced until 1947 when the Super A appeared, a model which did not grace the UK market.

Like the F20 and F30 tractors they replaced, the Farmall models H and M were very similar in appearance, and indeed featured the same wheelbase. The H had cylinder bore and stroke of 3.375 X 4.25 inches whilst the M had a 3.825 X 4.25 inch bore and stroke. The H engine was governed at 1.650 whilst the M ran at 1.450. Five forward speeds were provided, the top gear being blanked out when steel wheels were fitted. The usual Domestic (US) front wheel equipment was vee twins, but in the UK the wide front axle was more common.

A petrol start diesel model, the MD was also offered which had a similar engine to the M but this model did not venture across the Atlantic.

The new range of engines dispensed with the ball bearing crankshaft and were available with electric start. A multiplicity of extra equipment was on offer.

The rear axle track could be adjusted from 44 inch to 80 inch on the H and 52 inch to 88 inch on the M. Independant brakes were fitted as standard, but the little mudguards were optional.

New Standard Models

The W4 utilised the same engine and many other components as the Farmall H, but came out the year after that model in 1940. Like the new Farmall range it featured a cast frame and the actual transmission layout was very similar with a five speed gearbox again with top gear blanked out when steels were specified. Like the Farmall models, independant brakes acting on the different shaft could be used separately or together.

A number of W4 tractors found their way into Britain under "Lease-lend" and most of these appeared on steels, there subsequently being a fair number fitted with various types of pneumatic equipment after the war.

The W4 was produced until 1953 but ceased to be offered in the UK sometime before that.

The W6 shared the same engine and other components with the Farmall "M" and was to be seen in the UK in greater numbers than the W4 or W9 models. The WD6 diesel model was not imported, however, although the regular model was sold up to 1953, when Doncaster had commenced production of its own version.

Its popularity was limited post-war with the advent of hydraulics and three point linkage in the UK, which were not available for this particular model.

New Crawlers

W6 and W9 wheeled tractors and T6 and T9 crawlers also found their way across the Atlantic. It is unusual that very few, if any of the wheeled diesels came in, but the TD6 and TD9 crawlers did.

The early postwar scene found the "Farmall" range most predominant in 1946-47 sales along with the crawler range. Although the diesel variants of the wheeled tractors were featured in some sales literature, unlike the pre-war scene few, if any, were sold.

Whilst many International TracTracTors found use in agriculture their use in the construction field was also prominent, especially in later years with the introduction of the new models in 1939-41. These replace the T20, 35 and 40 range and were the TD6, TD9 which shared engine features with the WD6 and WD9 models and their VO counterparts the T6 and T9 which used the same engines as the W6 and W9.

All IH crawlers in the medium power area retained the petrol start diesel feature until the mid 'fifties when electric starting had been used in the UK for some time on the BTD6 model - the type of engine phased-in in the USA also.

The TD14 and T14 models were produced from 1939 to 1949. The T14 featured a 4-cylinder 4.75 X 6.5 inch engine which gave 60hp at 1,350rpm but was only produced for a short time in 1946. The TD14, which featured a diesel engine of the same bore and stroke gave 61hp at the belt, whilst the improved TD14A produced from 1949-1955 gave 75hp at the belt - but at 1.400 rpm.

The T9 and TD9 crawlers used the same engines as the W9 and WD9 wheeled models and both found their way into the UK. The most popular models in Britain were the TD6 and T6, leading to the British crawlers range introduced in the 1950s, the largest of which used the Rolls-Royce engine.

Details of all International engines are given in the table at Appendix 1.

The "Cub"

Cultivision, as it was known became a popular feature of the small IH tractors and lasted well into the 'fifties. With the full range of implements offered for the Farmall range it was possible to control many operations both mechanically and visibly from the tractor seat.

A smaller tractor, the "Cub" built with the same feature but with a smaller engine was introduced in 1947 and was produced until 1964. A number of these models saw importation into Southern Ireland and some of the later ones have reached the shores of mainland Britain through preservation.

The Farmall BM

The first British-built Ms were available from International's Doncaster factory by September 1949. These tractors differed from their US counterparts by the use of an improved front axle mounting with the track rod fitted behind and under the frame, this also made the tractor some 18 inches longer.

The four cylinder engine gave outstanding fuel economy and used a Tocco-hardened crankshaft. Great attention was also taken to ensure that oil, air and fuel filtering accessories were fitted to protect against dirt and impurities.

A combination manifold with intake heat control was used and the radiator shutter was adjustable from the tractor seat in conjunction with the bonnet mounted heat indicator. Standard wheel equipment was 11" X 38" rear tyres, and 600" X 16" front. Both axles were adjustable giving 57" - 81" at the front and 52" - 88" at the rear. A swinging drawbar was fitted as standard, but the hydraulic lift-all was listed as additional equipment.

Tractors supplied below serial no. D2342 with starting and lighting attachments were fitted with Delco-Remy components; tractors with serial no. 2342 upwards, had Lucas units.

By the early 1950s, the "M" transfer on the engine cover had been replaced with a "BM" motif, prior to the introduction of the diesel range.

The BMD and BWD6

1952 saw the introduction of the diesel engined BMD, which was upgraded the following year as the Super BMD, but still retained the lift-all hydraulic system. In 1954, at the Royal Show, an entirely new tractor was exhibited by International, this was the SBWD6, designed for farmers who required a tractor with the same power as the SBMD, but without the rowcrop features special to that model.

Both tractors used the Doncaster-built 50hp BD264 engine which was a true diesel, eliminating the complicated petrol starting of the American machines. Glow plugs in the combustion chambers ensured good starting from cold and the automatic torque control in the governor gave steady engine speed under all load conditions. As can be seen from the illustrations, the BWD6 shared many components with the Farmall, mainly the engine, side channels, transmission casting and fuel tank. The BWD6 also boasted a "live" hydraulic system, consisting of a gear type pump mounted at the front of the engine, which came into operation the moment the engine was started, eliminating the use of the clutch when turning at the end with mounted implements. A special point, worth noting here, was the fact that the drawbar was fitted to the linkage and rode up-and-down with the arms making coupling to trailed implements very easy.

A petrol start TVO model of these tractors was also available, called the Farmall Super BM and the Super BW6.

The Super BMD model was retained for a number of years and the last lot made inherited the BWD6 type lift necessitating in the battery holder being moved under the seat, this tractor also had disc brakes and the headlights were mounted on the side of the radiator cowl.

The B250 and B275

In the mid-1950s, many firms were pushing on with designing a smaller tractor to combat the share of the market held by Ferguson and IH, who had purchased the old Jowett factory, at Bradford, announced the first British designed and built tractor from that factory in 1956 in the form of the B250.

This light, diesel tractor used the BD144 engine, providing 30hp, with a 5-forward, 1-reverse gearbox, disc brakes and differential lock, fitted as standard, but the hydraulic lift was fitted as an extra. Wheel equipment was 10.00 X 28" rears and 4.00 X 19" fronts, as standard but, 11.00 X 28" rears and 5.50 X 16" fronts could be factory fitted, at extra cost.

The B250 was designed with the drivers needs in mind, wide mudguards and foot boards were fitted, along with the uncluttered engine cover.

Now that Fordson and Massey Harris had joined the lighter-tractor range with the Dexta and MF35 with dual range gearboxes, International announced their next model.

Introduced in 1958, along with the B450, the new light tractor was called the B275 and was, in effect, a B250 with 10-speed gearbox and beefed-up engine giving 35hp.

This was a general-purpose model with mechanical depth control hydraulics with Category 1 and 2 three point linkage.

The B275 Hi-clear had the same features as the standard model, but offered extra big clearance under the front axle and used larger rear wheels.

Introduced in 1958, the large B450 was a more refined BWD6 under new tinwork with an inbuilt hydraulic power lift, differential lock and disc brakes. The IH BD264 engine had been improved to give 55hp by this time, and the B450 was capable of pulling a five furrow mounted plough. This tractor was available with 11" X 38" rear tyres and 600" X 19" fronts, and adjustable axle, or, 13" X 30" rears with 7.50" X 16" fronts on a fixed axle but, in 1959, a Rowcrop version was introduced, with variable tread front wheels fitted as standard. As can be seen in the illustrations, the new rowcrop model looked just like the Super BMD front and grated onto a B450 back, and which, in fact, is what it was. The Farmall M type steering was used on the rowcrop and now that the inbuilt hydraulic unit was fitted, the battery tray was placed back under the fuel tank.

A choice of axles were also available, under alternative equipment, for rowcrop work.

After 2-years of severe testing, a 4-wheel-drive B450 was added to the range, this was built in conjunction with Roadless Traction and gave a first-class, powerful tractor, with excellent wheel grip, when needed.

More Crawlers

The TD14 and T14 models were produced from 1939 and 1949. The T14 featured a 4-cylinder, 4.75" X 6.5" engine, which gave 60hp at 1,350rpm, but was only produced for a short time in 1946. The TD14, which featured a diesel engine of the same bore and stroke, gave 61hp at the belt, while the improved TD14A - produced from 1949-55 - gave 75hp at the belt, but at 1,400rpm.

The T9 and TD9 crawlers used the same engines as the W9 and WD9 wheeled models and both found their way into the UK. The most popular models in Britain were the TD6 and T6, leading to the British crawler range, introduced in the 1950s, the largest of which use the Rolls-Royce engine.

IH had also built crawlers at Doncaster, starting with the BTD6 in 1953. The first two years production were 40hp diesel-engined machines but, in 1955, these were improved to give 50hp and were joined by a BT6 model.

In 1957, a new crawler, the BTD640, was introduced for agricultural use, this also used the BD264 engine, with lighter track equipment than the BTD6. After a production run of 6-years, a new BTD5, 40hp crawler appeared, which had a planetary steering system, 40" or 48" gauge, four of five roller track frames, Category 2, three-point linkage, eight forward and two reverse speed gearbox and fully "live" hydraulics.

Models up-to-1970

July 1961 saw the introduction of the 436hp B414 tractor, with its distinctive white wheels and grill panels, followed in 1964 by the similarly styled 65.5bhp, 614 model. The B414 was in production until February 1966, when it was replaced by the 434. July 1968 saw the B614 replaced by the 66bhp 634. This was not the end of the B-designation, as the faithful B450 was built in both 2 and 4-wheel-drive form until 1970.

A line-up of IH products outside the factory.

Below: Harvester House, 259 City Road, London; the British headquarters of the IH company.

The Mogul 8/16 was introduced in 1914 and was produced until 1917 when it was replaced by a somewhat larger 10/20 model. It featured a horizontal single cylinder engine of 8" bore by 12" stroke which gave 16hp at 400rpm. The engine was designed to burn gasoline or kerosine, was hopper cooled and featured make and break ignition, supplied by an oscillating magneto. A compression relief cam aided starting. Oiling was by automatic force feed system, and a centifugal governor was fitted. Steering was by worm and sector mounted over the front carriage. The transmission was of an unusual design, being of the epicyclic variety, but only a single forward speed was provided. Final drive to the differential was by roller chain.

The Mogul 12/35 was originally introduced as the 10/20 in 1911, but renamed in 1913 and continued in production until 1918. A small number of this model found their way into the UK and these were put to good use by the Ministry of Food Production. The Mogul 12/25 featured a horizontally opposed two cylinder engine of 7" bore by 8" stroke developing 25hp at 550rpm. Two forward and one reverse speed were fitted. Lubrication and ignition arrangements were similar to the 8/16, but starting was aided by a friction wheel bearing on the flywheel. Ackerman steering was fitted, however.

Opposite page. Top: The Titan 10/20 is possibly one of the best known of the early tractors in the UK. With its twin cylinder engine of 6+" bore by 8" stroke, ignition was provided by rotary high tension magneto, and a centifugal governor was fitted, but an unusual feature was that both cylinders operated in tandem, there being a power stroke from one or the other on every revolution. The engine speed of 500rpm was later increased to 550rpm. A simple fuel mixer finally gave way to a proper carburettor in 1921.

Centre: The view of the engine shows clearly the force feed lubrication system with it oiler on top of the cylinder block.

Bottom: A Titan at work with one of IHCs own 'Hamilton' tractor ploughs.

TITAN
MADE BY
INTERNATIONAL HARVESTER
CORPORATION
MILWAUKEE WORKS
20 H.P.

TITAN TRACTOR
W.C. BRIDGES
CIRENCESTER

Top: The first 'Junior' tractors carried the type designation VB; production started in 1917. No aircleaner was provided, purely a cowl on the end of the intake pipe.

Centre: The improved HC series was introduced in 1919 and featured modifications to the engine crankcase plus the provision of a water air washer. The engine was identical to that then fitted to IHCs 'G' series trucks and indeed the radiator behind the engine was copied from this layout which was a common feature of many early automobiles.

Bottom: The final series of 8/16 was the IC which was only produced until 1922. It featured a repositioned exhaust, aircleaner and additional tinwork to support the fuel tank. In 1919 the 'Junior' was sold in the UK for £325, although the final price seems to have been as low as £275, by 1922.

Opposite page, top: The original 10/20 came in 1923 and replaced the Titan 10/20. Note the 'Junior' designation on this early example.

Centre: The 15/20 came in 1921 but this 1934 example differed little in appearance from the originals, save for the oilbath cleaner, restyled bonnet side panels, and the provision of a more powerful engine.

Lower: A 1934 10/20 complete with spade lugs and nearside fender extension. At this time the 10/20 sold for £230 on steels and £266 on rubbers, whilst the 15/30 sold for £320 in 1930.

Bottom: A partially dismantled 15/30 hull showing the engine in position.

INTERNATIONAL
JUNIOR

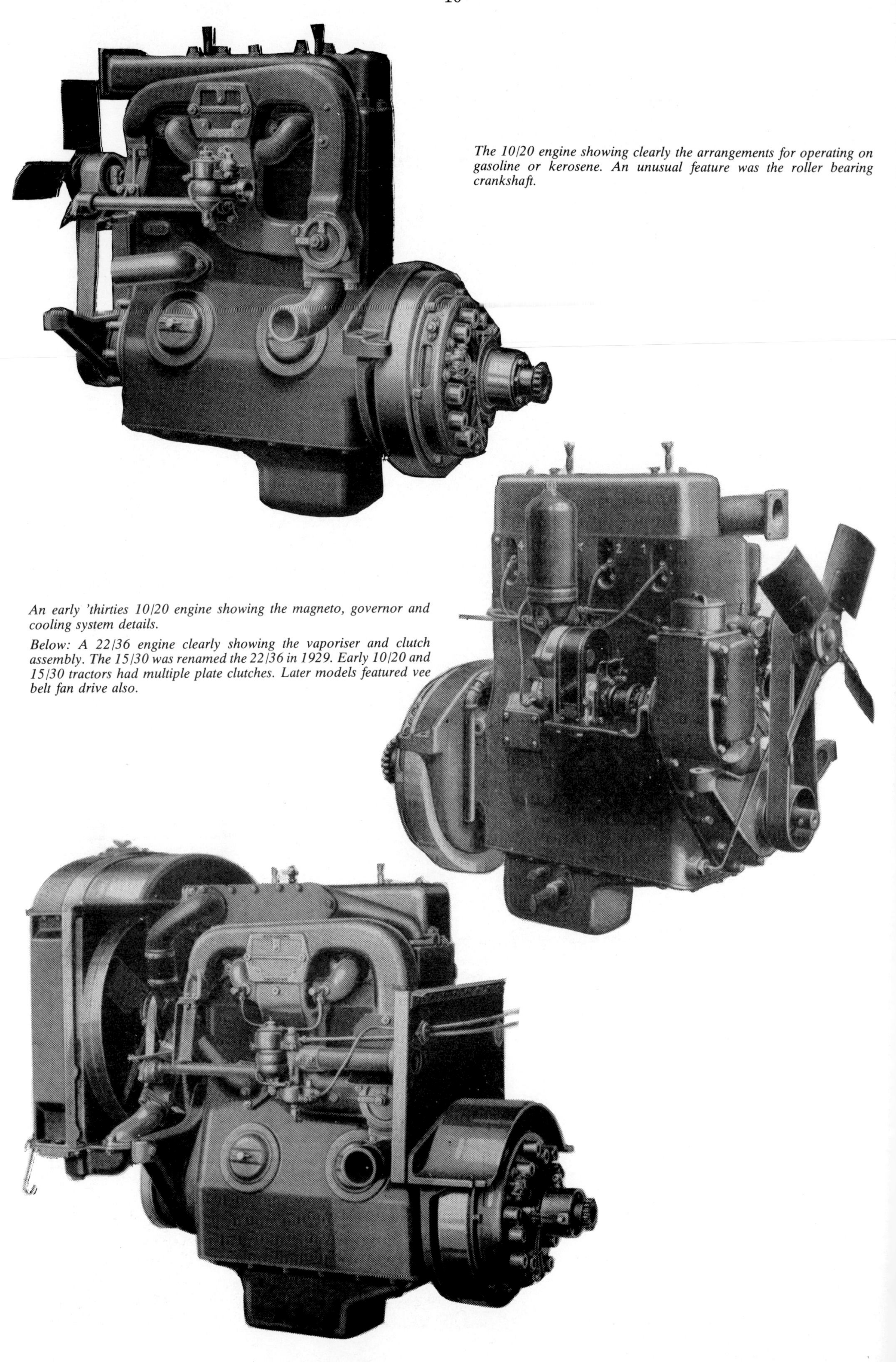

The 10/20 engine showing clearly the arrangements for operating on gasoline or kerosene. An unusual feature was the roller bearing crankshaft.

An early 'thirties 10/20 engine showing the magneto, governor and cooling system details.

Below: A 22/36 engine clearly showing the vaporiser and clutch assembly. The 15/30 was renamed the 22/36 in 1929. Early 10/20 and 15/30 tractors had multiple plate clutches. Later models featured vee belt fan drive also.

Orchard fenders were available for the 10/20.

A 10/20 with special rear wheels and road lugs.

A 10/20 on pneumatics with French & Hecht centres as usually seen in the USA.

For UK consumption, it was usual to find Dunlop cast centres fitted. Removal of the engine and gearbox did not involve splitting as on other tractor designs.

Two pages of early IHC tractors at work. The decals could show International McCormick Deering or just Deering, in some territories. The two shots on this page show such detail with two 22/36 tractors.

Opposite page, top: A delightful harvesting scene with the International providing power for the stationary baler, and a contemporary Fordson on the carting.

Centre, and bottom: A late 10/20 at work with an Allis Chalmers combine. Note the windmill in the centre of the first picture.

A W30 with radiator chaff guard, manifold shield and the later type of fenders.

Detail differences can be seen on this earlier example of a W30. Note the use of a radiator blind instead of the later shutter.

Steel wheel equipment was standard on the W30 but, in the UK, Dunlop wheels and centres were available. The W30 cost £285 on steels and £321 on rubbers.

Engine side panels were not usual on the W30, nor were the French & Hecht spoked wheel centres seen much in the UK. The W30 was also available in industrial form as the I30 (below).

The W40 was introduced in 1934. The example, above, features French & hecht wheels, which were not offered in the UK on this model. The WK40 (right) shows the normal steel wheel equipment and vaporiser for running on paraffin. The six cylinder engine gave a very smooth running machine.

An I40 was available, also for industrial use. A four speed gearbox was fitted.

The WD40 featured International's own petrol start diesel engine, and was the first diesel wheeled tractor to be offered in the USA. Like the W40, a three speed transmission was offered, but the ratios were difficult to account for the slower engine speed of the diesel.

The ID40 was offered for commercial use, and had a special gearbox with an extra top speed.

At the other end of the scale was the W12, which was built from 1934-38. The mechanical units were shared with the F12, and the tractor was available on steels or rubbers. From 1938 through 1939, the tractor, with engine speed increased from 1400 to 1650rpm, was sold as the W14. This W12,(right), would have cost £195 in 1936 on steel wheels, and £235 on pneumatics, as shown below.

A W14 on pneumatic tyres.

This is a W14 on steels – note the addition of a radiator shutter.

A Fairway 14 with special wide wheels to give low ground pressure. It also had a special high gear of 10$\frac{3}{8}$mph, not found on the W14, on steel wheels.

An O14 for orchard use.

The Farmall regular with normal vee twin front wheels.

Above: The Farmall regular became the F20 in 1932 and gained an extra forward speed. This is a wide axled variant.

Below: The larger Farmall 30, easily identified by the style of wheels, and the shape of the gearbox casting.

The left side of the Farmall 30. The manifold arrangements and the fact that the engine has a water pump, easily identify this model from the F20.

Below: An F20, fitted with rear mounted toolbar and ridger.

Bottom: A mid-1930s shot of International equipment being demonstrated in Lincolnshire.

Above: The F12 with standard wheel equipment and wide front axle.

Right: Dunlop equipment on the F12. From 1936, all Farmall models adopted the new red colour scheme.

F12 with single front wheel (left). French & Hecht wheel centres, plus optional fenders, are provided (below).

The F14 featured a higher driving position, and the wide axle variety is seen above. Pneumatics were popular with this model, and such a tractor is seen below.

The T20 TracTracTor. The view above shows clearly the covers for the steering clutches and brakes above the drawbar. A T20 would cost you £450 in 1937 and, in all, 440 units were imported into the UK. The view to the left shows the aircleaner position.

A T20 for orchard work with lower driving position is seen below.

The T35 (above) and TD35 crawler (below) were produced from 1936-39, but only 45 units came to Britain. They were similar to the TD40 and T40 units but had an eighth of an inch less in the bore on the T35 and a quarter of an inch less on the TD35. Five forward speeds were provided, and when the T35 was Nebraska tested, it gave 35.9 drawbar horsepower, whilst the TD35 gave 42.4 under similar test conditions.

The TD40 (above) and T40 crawlers were the largest of the old style TracTracTors. The same six cylinder engine as the W40 was used in the T40, whilst International's petrol start diesel was used in the TD40.

The rear view of the T40 (above) shows the similar transmission layout to the T20. The six cylinder engine used in the W40 and T40 tractors is seen to the left. It was 3¾" bore x 4¼" stroke.

The TD14 was typical of the larger IH Crawlers imported into the UK and paved the way for the more comprehensive range of domestically built models in the 1940s and 1950s. It was the first of the styled crawlers launched in 1939.

The Farmall H (right) was phased in during 1939 but did not appear in the UK until 1940-41, and most of the wartime examples were fitted with steel wheels, as seen below, complete with toolbar.

The larger Farmall M became a familiar sight on British farms. Note the axle layout of US built examples.

Not seen in the UK was this MV variant for vineyard use.

The compact power unit of the Farmall M was also used in the W6 and T6 models. Gone were the old features, such as roller bearings on the crankshaft. These units gained a reputation for longevity and good performance, and were easy to overhaul when the time came.

Below: The Farmall M was often sold in the USA in more usual vee front wheeled form.

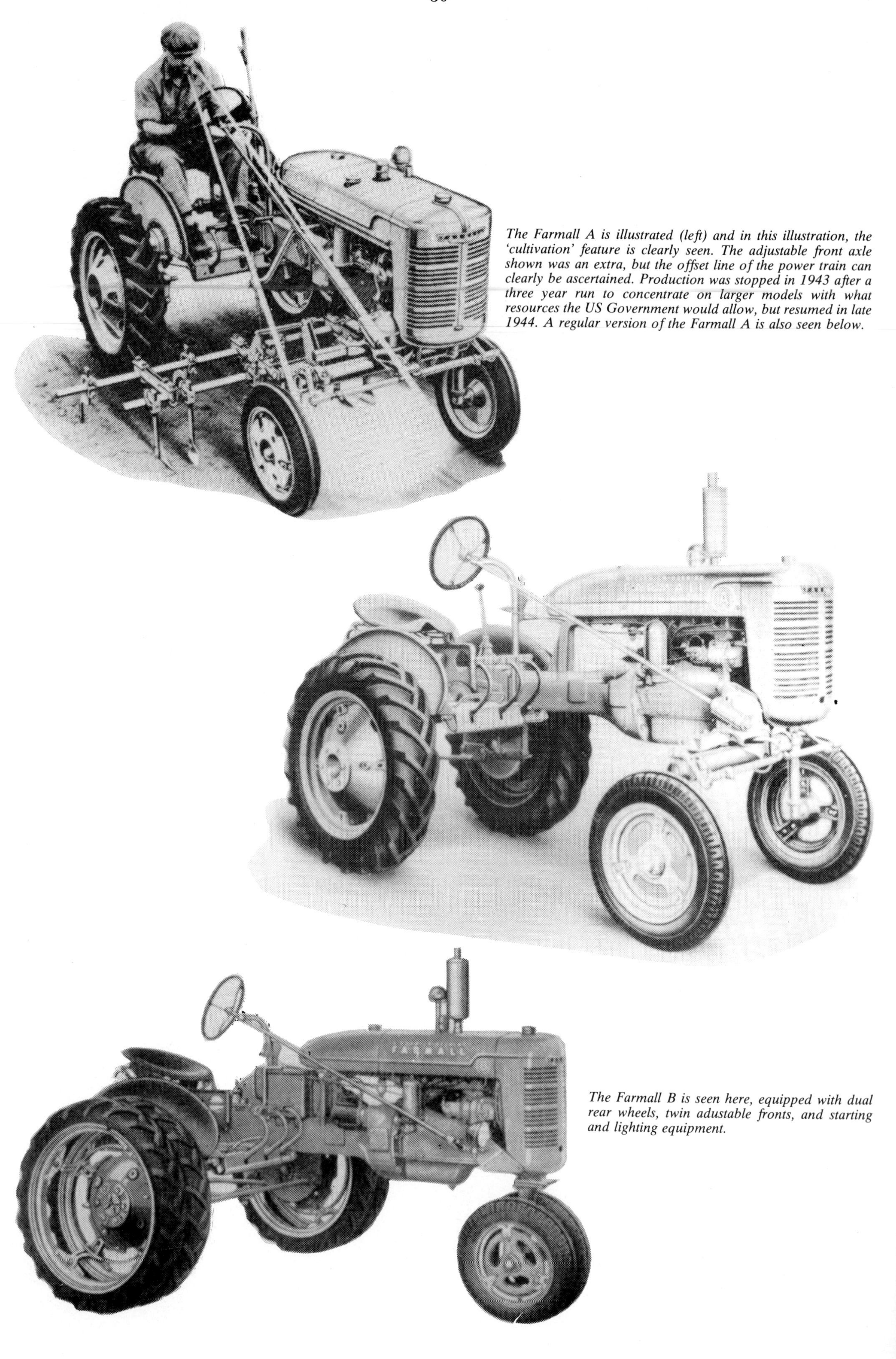

The Farmall A is illustrated (left) and in this illustration, the 'cultivation' feature is clearly seen. The adjustable front axle shown was an extra, but the offset line of the power train can clearly be ascertained. Production was stopped in 1943 after a three year run to concentrate on larger models with what resources the US Government would allow, but resumed in late 1944. A regular version of the Farmall A is also seen below.

The Farmall B is seen here, equipped with dual rear wheels, twin adustable fronts, and starting and lighting equipment.

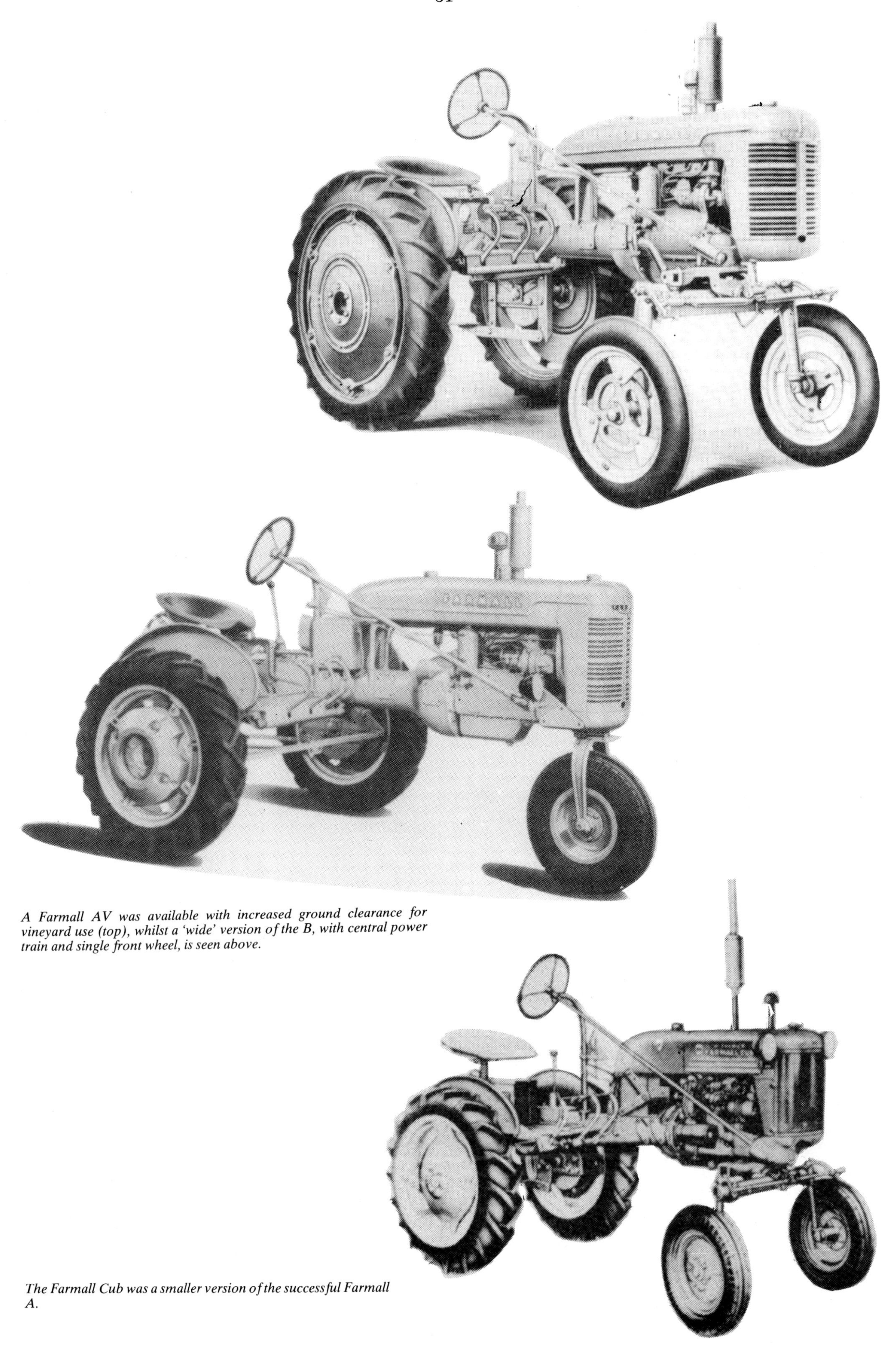

A Farmall AV was available with increased ground clearance for vineyard use (top), whilst a 'wide' version of the B, with central power train and single front wheel, is seen above.

The Farmall Cub was a smaller version of the successful Farmall A.

The W4 used the same engine as the Farmall H and was the smallest of the 'standard' models. Pneumatic tyre equipment is seen on the example to the left.

More common in wartime was the steel wheeled W4.

The OS4 was a 1945 introduction, which was an economy version of the O4 orchard model, lacking the special fenders which gave the original tractor a similar appearance to the O6, opposite.

The W6 was the next in size of the standard models and is seen here, on steel wheels.

Using the same engine as the Farmall M, the W6 is seen here on pneumatics.

The O6 featured citrus fenders for orchard and vineyard use.

The W9 on steel wheels.

The W9 on pneumatics.

The WR9 rice-field, with special rear fenders and oversize rear tyres.

A WD9 with International's own petrol start diesel. A few of these did reach the UK, although there is the story that one boat load, which contained most of the diesels, was torpedoed.

Below, and bottom: W9s at work.

McCORMICK
INTERNATIONAL
FARMALL

FARMALL
M
TRACTOR
No1
THE FIRST DONCASTER
INTERNATIONAL
FARMALL M

A British Farmall, showing the new style front axle mounting. Also shown is the position of the battery box and field lighting set.

Below: A much later BM, by this time the battery had been moved to under the seat, to accommodate the power lift unit, and the headlights are now positioned at the front of the tractor.

Opposite page, top: The first British-built Farmall M, seen coming off the line, at the new IH factory, in Doncaster. Waiting to take the wheel is the Minister for Agriculture, seen seated on the left.

Bottom: The new tractor awaits despatch.

Introduced in 1952, the Super BMD was fitted with the company's own Doncaster-built BD264 engine, but retained the 'lift all' hydraulic system. The shot on the right shows an earlier tractor.

Below: A late SBMD, showing the new 'live' hydraulic system, with the oil reservoir positioned under the fuel tank. The two six-volt batteries were now fitted under the seat, and the headlights moved to the sides of the front cowl.

Opposite page: The 1,000th BMD tractor to leave Doncaster did so on 16 July, 1953, and was painted in a gold livery. The same tractor is seen ploughing, in the lower photo.

The new SBW6 tractors were the standard versions of the Farmall range, and offered the new 'live' hydraulic system, driven from a direct connected gear type pump. The pipe work of which can be clearly seen passing under the belt pulley. The shot below, shows this model without hydraulics; an adjustable swinging drawbar would be fitted.

Opposite page, top: The Super BWD6 is seen here, with mounted 2-furrow reversible plough. The double acting cylinder, which powers the lift up and down, can be seen next to the LH mudguard.

Bottom: I.H. tractors leave Doncaster by rail, minus tyres, and other parts crated and fixed to the drawbar. This consignment, hauled by an ex-Great Northern J2 class 0-6-0 locomotive, is destined for Australia, where import restriction prevented tractors being shipped in with rubber tyres.

McCORMICK
INTERNATIONAL
STANDARD

Two views of the Super BW6 tractor. Wheel adjustment was by changing the position of the rims on the wheels to give four front settings, and by moving shaped clamps, three rear wheel settings could be obtained.

British-built BTD6 tractors followed closely the style of the TD6 but used the British-built BD264 engine. The tractor in the lower view is an industrial model with long tracks.

Right, and centre: Designed and built in Britain, the new B250 tractor was introduced in 1956.

Bottom: A petrol engine was available for the B275, but not until 1961. This was three years after the diesel model was offered.

Top: The B275. Centre: A high clearance B275. Bottom: The industrial version came in during 1959. B275 units were used by Steelfab and Whitlock for digger / loaders.

The B450, showing alternative wheel equipment. Note the adoption of the standard 'Sankey' front and rear wheel pressings on this example.

Another B450, this time with adjustable front wheels.

Introduced in 1959, the rowcrop versions (RC) of the B450 is shown here with vee frontwheel equipment. Note styling, as previous BM and Super BMD models.

Opposite page, top: The first B450 diesel is about to leave the works in October 1958. Bottom: A B450 Farmall, with 19" front and 14x30" rear tyre equipment.

NO
SMOKING
1st
B 450
DONCASTER WORKS
OCTOBER 1958
B-450

McCORMICK
INTERNATIONAL
DIESEL
FARMALL
B-450
Exide
Exide

A B450 at work, with an IH 313 three furrow plough.
Below: From 1964, the B450 four-wheel drive was available and used an axle, built in conjunction with Roadless Traction.

Above: A B450 seen here operating a B55T PTO driven baler, the B55T could also be had with its own diesel engine. It could bale up to 10-tons per hour.

The BTD-640 was introduced in 1957, and featured lighter track equipment than its predecessor, the BTD-6.

McCORMICK
INTERNATIONAL
B414

INTERNATIONAL
B414
48·GLO

Opposite page, top: The new B414 43bhp tractor, introduced in July 1961. This model had an 8-speed forward gearbox. This example is fitted with a deluxe seat and power steering.

Bottom: An impressive view of the B614 62.5bhp tractor, which was available in early 1964. This used the BD281 engine, with an eight forward-2 reverse speed gearbox.

This page: First seen in November 1967, the B614 4-wheel drive was only made for a short period – until July 1968 – before being replaced by the 634 66.6bhp model.

Below: The B434 replaced the B414, and had new styling. Here, a B414, masquerading in the new tinwork prior to the introduction of the new model.

The AOS6 Tractor was the Australian cousin of the BWD6 but featured some items borrowed from the OS-6 built in the United States. It had a reduced engine bore of 3.875" giving 29/32HP and had three point linkage operated through the 'Liftall' hydraulic system.

The Super AW6 was the Australian equivalent of the BW6 but did not have a live hydraulic system. The engine fitted was exclusive to Australia being classified AC-264, being designed to operate on Australian spec. fuels, and a six rather than a twelve volt electrical system was used.

SERIAL NUMBERS.

All 8-16, 10-20, and 12-25 Mogul tractors have engine number stamped on R.H. end of crankshaft. All 8-16 and 10-20 tractors have number stamped on front bolster. 12-25 models have the tractors number stamped on front main frame channel.

TITAN TANK COOLED TRACTORS
12,15,20,25,18-35,45 and 30-60 engines.
Friction Drive, 1908-09, T5000 to T5236.

Type C Mogul, 20 HP, Prefix TL 1909-14
Type C Mogul, 25 HP, Prefix TP 1910-14
Type D Titan, 20 HP, Prefix TD 1910-14
Type D Titan, 25 HP, Prefix TM 1910-14
Type D Titan, 18-35 HP, Prefix TB & TC 1913
Type D Titan, 18-35 HP, Prefix TF 1914-16
Type D Titan 27-45 HP, Prefix TN 1911
Type D Titan 27-45 HP, Prefix TA 1912
Type D Titan 27-45 HP, Prefix TH 1913
Type D Titan 27-45 HP, Prefix TT 1913
Type D Titan 27-45 HP, Prefix TR 1914
Type D Titan 30-60 HP, Prefix TK 1914-17
Type D Titan 30-60 HP, Prefix TJ 1914-17

INTERNATIONAL 15-30 HP KEROSENE TRACTOR (WAS TITAN)

TS101 to TS344, Titan "L" head, ... 1915
TS345 to TS883, Titan "L" head, ... 1916
TS884 to TS918, Titan "L" head, ... 1917
TW101 to TW896, Titan, redesigned, ... 1917
EC501 to EC1000, International w/cab, ... 1918
EC1001 to EC1365, International w/o cab, ... 1918
EC1366 to EC3017, International w/o cab, ... 1919
EC3018 to EC4085, Internation w/o cab, ... 1920
EC4086 to EC4910, International w/o cab, . 1921-22

IHC MOGUL TRACTORS

20 HP, K501 to K510 on Type C trucks 1911
10-20 HP, B501 to B585 1913-1914
20 HP Mogul Jr., J501 to J1312 1911-13
15-30 Mogul (re-rated 25 HP) C1313 to C1840
.... 1913-15
45 HP, X501 to X550 1910
45 HP, X551 to X1081 1911
45 HP, X999 to X1080 and X1085 to X2173 1912
30-60 HP (re-rated 45 HP), U2174 to U29401913-15

IHC MOGUL TRACTORS

8-16 HP, SB501 to SB3750 1915
8-16 HP, SB3751 to SBN15000 1916-17
10-20 HP, BC501 to 1917-19
12-25 HP, F501 to F750, "L" head engine) .. 1913-18
12-25 HP, F751 to F2100, "Round" head engine)

McCORMICK-DEERING 15-30 GEAR DRIVE TRACTOR

PREFIX LETTERS "TG"
112 to 310 1921
311 to 1660 1922
1661 to 6546 1923
6547 to 13867 1924
13868 to 26845 1925
26846 to 48646 1926
48647 to 64400 1927
64401 to 99925 1928

NEW 15-30, LATER CALLED 22-36
99926 to 128236 1929
128237 to 150127 1930
150128 to 154507 1931
154508 to 156212 1932
156301 to 157477 1934

TITAN 10-20 HP, HORIZONTAL, 2-CYLINDER CHAIN DRIVE TRACTOR
Serial shop number stamped on front end of right hand channel. Tractors with "TV" prefix operate at 500 RPM, Tractors with "TY" prefix operate at 575 RPM.

TV116 to TV2356 1916
TV2357 to TV11397 1917
TV11398 to TV29072 1918
TV29073 to TV46306 1919
TV46307 to TV50235 1920
TY50236 to TY67810 1920
TY67811 to TY75539 1921
TY75540 to TY78464 1922

INTERNATIONAL 8-16 HP KEROSENE TRACTOR, 4-CYLINDER, CHAIN DRIVE

Serial number stamped on finished surface under magneto bracket. Engines with a "VB" prefix use a 3-bearing crankshaft and 4" bore and 5" stroke. The same bore and stroke is used in engines with an "HC" prefix, but they use a 2-bearing crankshaft. "IC" engines have a 4.5" bore x 5.5" stroke with a 2-bearing crank.

VB501 to VB3700 1017-1918
VB3701 to VB5748 1919
HC501 to HC6023 1919
HC6024 to HC11871 1920
IC501 to IC9734 1921
IC9735 to IC17023 1922

McCORMICK-DEERING 10-20 HP GEAR DRIVE TRACTOR

Prefix letters "KC" (Regular) and "NC" (Narrow Tread). After 1934 Narrow Tread was designated by the addition of an "NT" suffix to the Regular 10-20 Serial Number.

"KC" SERIAL NUMBERS

501-7640 1923
7641-18868 1924
18869 to 37727 1925
37728 to 62823 1926
62824 to 89469 1927
89470 to 119822 1928
119823 to 159110 1929
159111 to 191485 1930
191486 to 203174 1931
201013 to 204138 1932
204239 to 206178 1934
206179 to 207274 1935
207275 to 210234 1936
210235 to 212424 1937
212245 to 214885 1938
214886 to 215973 1939

"NT" SERIAL NUMBERS

501 to 648 1926
649 to 831 1927
832 to 1154 1928
1155 to 1542 1929
1543 to 1749 1930
1750 to 1832 1931
1833 to 1911 1932
1912 to 1951 1933
1952 to 1960 1934

SERIAL NUMBERS

FARMALL REGULAR AND FAIRWAY TRACTORS

Prefix Letters QC and T
PREFIX LETTERS "QC"

501-700	1924
701- to 1538	1925
1539 to 5957	1926

PREFIX LETTER "T"

5969 to 15470	1927
15471 to 40369	1928
40370 to 75690	1929
765691 to 117783	1930
117784 to 131871	1931
131872 to 134954	1932

FARMALL F-30 TRACTORS

PREFIX LETTERS "FB"

501 to 1183	1931
1184 to 4304	1932
4305 to 5525	1933
5526 to 7031	1934
7032 to 10406	1935
10407 to 18683	1936
18684 to 26848	1937
27186 to 28719	1938
29007 to 30026	1939

FARMALL F-12 TRACTORS

PREFIX LETTERS "FS"

501 to 525	1932
526 to 4880	1933
4881 to 17410	1934
17411 to 48659	1935
48660 to 81836	1936
81837 to 117517	1937
117518 to 123942	1938

FARMALL F-14 TRACTORS

PREFIX LETTERS "FS"

124000 to 139606	1938
139607 to 155902	1939

FARMALL F-20 TRACTORS

Prefix Letters "FA" and "TA". "FA" letters through No. 6381, with "TA" letters following.

501 to 3000	1932
3001 to 6381	1934
135000 to 135661	1934
6382 to 32715	1935
32716 to 68748	1936
68749 to 105596	1937
105597 to 130864	1938
130865 to 134999	1939
135700 to 148810	1939

McCORMICK-DEERING W-12 TRACTOR

PREFIX LETTER "WS"

503 to 1355	1934
1356 to 2030	1935
2031 to 2767	1936
2768 to 3798	1937
3799 to 4133	1938

McCORMICK-DEERING W-14 TRACTORS

PREFIX LETTERS "WS"

4134 to 4609	1938
4610 to 5296	1939

McCORMICK-DEERING 0-12, 0-14, FAIRWAY 12, 14 TRACTORS

PREFIX LETTERS "OS" OR "FOS"

512 to 1091	1934
1092 to 1625	1935
1626 to 2276	1936
2277 to 3260	1937
3261 to 3881	1938
3882 to 4287	1939

McCORMICK-DEERING W-30 TRACTORS

PREFIX LETTERS "WB"

501 to 521	1932
522 to 547	1933
548 to 3181	1934
3182 to 9722	1935
9723 to 15094	1936
15095 to 23833	1937
23834 to 29921	1938
29922 to 32481	1939
32482 to 33041	1940

McCORMICK-DEERING W-40 AND WD-40 TRACTORS

PREFIX LETTERS "WAC", "WKC", OR "WDC".

501 to 1440	1935
1441 to 5119	1936
5120 to 7664	1937
7665 to 9755	1938
9756 to 10322	1939
10323 to 10559	1940

FARMALL A,AV,B,BN TRACTORS

PREFIX LETTERS "FAA", "FAAV", "FAB", "FABN", "1AA".

501 to 6743	1939
6744 to 41499	1940
41500 to 80738	1941
80739 to 96389	1942
96390 to 113217	1944
113218 to 146699	1945
146700 to 182693	1946
182964 to 198298	1947
200001 to 220829	1947

FARMALL H AND HV TRACTORS

PREFIX LETTERS "FBH" AND "FBHV"

501 to 10652	1939
10653 to 52386	1940
52387 to 93236	1941
93237 to 122589	1942
122590 to 150250	1943
150251 to 186122	1944
186123 to 214819	1945
214820 to 241142	1946
241143 to 268990	1947
268991 to 300875	1948
300876 to 327974	1949
327975 to 351922	1950
351923 to 375860	1951
375861 to 390499	1952
390500 to 391730	1953

FARMALL M, MV, MD, MDV TRACTORS

PREFIX LETTERS "FBK", "FBKV", "FDBK", "FDBKV"

501 to 7239	1939
7240 to 25370	1940
25371 to 50987	1941
50988 to 60010	1942
60011 to 67423	1943
67424 to 88084	1944
88085 to 105563	1945
105564 to 122822	1946
122823 to 151707	1947
151708 to 180513	1948
180514 to 213578	1949
213579 to 247517	1950
247518 to 290922	1951
290923 to 298218	1952

FARMALL CUB TRACTOR

PREFIX LETTERS "FCUB"
501 to 11347 1947
11348 to 57830 1948
57831 to 99535 1949
99536 to 121453 1950
121454 to 144454 1951
144455 to 162283 1952
162284 to 179411 1953
179412 to 186440 1954

FARMALL C TRACTOR

PREFIX LETTERS "FC"
501 to 22623 1948
22524 to 47009 1949
47010 to 71879 1950
71880 to 80432 1951

McCORMICK-DEERING 0-4, OS-4, W-4 TRACTORS

PREFIX LETTERS "OBH", "OBHS", "WBH".
501 to 942 1940
943 to 4055 1941
4056 to 5692 1942
5693 to 7592 1943
7593 to 11170 1944
11171 to 13933 1945
13934 to 16021 1946
16022 to 18879 1947
18880 to 21911 1948
21912 to 24469 1949
24470 to 28166 1950
28167 to 31213 1951
31214 to 33066 1952
33067 to 1953

McCORMICK-DEERING O-6, OS-6, ODS-6, W-6, WD-6 TRACTORS

PREFIX LETTERS "OBK", "OBKS", "ODBKS", "WBK", "WDBK".
501 to 1224 1940
1225 to 3717 1941
3718 to 5056 1942
5057 to 6312 1943
6313 to 9495 1944
9496 to 14152 1945
14153 to 17791 1946
17792 to 22980 1947
22981 to 28703 1948
28704 to 33697 1949
33698 to 38517 1950
38518 to 44317 1951
44318 to 45273 1952
45274 to 1953

McCORMICK-DEERING W-9, WR-9, WD-9, WDR-9 TRACTORS

PREFIX LETTERS "WCB", "RCB", "WDCB", "RDCB".
501 to 577 1940
578 to 2992 1941
2993 to 3650 1942
3651 to 5393 1943
5394 to 11458 1944
11459 to 17288 1945
17289 to 22713 1946
22714 to 29206 1947
29207 to 36158 1948
36159 to 45550 1949
45551 to 51738 1950
51739 to 59406 1951
59407 to 64013 1952
64014 to 67919 1953

10-20 TRACTRACTOR

PRFIX LETTERS "TT"
501 to 700 1928
701 to 975 1929
976 to 1527 1930
1528 to 2005 1931

T-20 TRACTRACTOR

PREFIX LETTERS "ST"
501 to 550 1931
551 to 2052 1932
2053 to 2527 1933
2528 to 3275 1934
3276 to 5386 1935
5387 to 8500 1936
8501 to 12517 1937
12518 to 14549 1938
14550 to 15699 1939

TA-40 AND TD-40 TRACTRACTORS

PREFIX LETTERS "TAC", "TCC".
501 to 732 1932
733 to 1246 1933
1247 to 1813 1934

T-40 AND TD-40 TRACTRACTORS

PREFIX LETTERS "TAC", "TKC", "TCC", "TDC".
2501 to 2784 1934
2785 to 4392 1935
4393 to 6302 1936
6303 to 7718 1937
7719 to 8598 1938
8599 to 9565 1939

T-35 AND TD-35 TRACTRACTORS

PREFIX LETTERS "TKB", "TDKB".
507 to 2969 1937
2970 to 4628 1938
4629 to 6092 1939

The BD264 engine

INTERNATIONAL HARVESTER GREAT BRITAIN SERIAL NUMBERS.

M&BM	Built
D1001-D1258	1949
D1259-D3090	1950
D3091-D5000	1951
B5001-B6000	1951
BF6001-BF6210	1951
BFC6211-BFC6589	1951
BFC6590-BFC9484	1952
BFC9485-BFC10431	1953
BFC10432-BFC10846	1954

BMD	
1952	504-532
1953	533-1432

SBM	
5001-5849	1953
5850-7764	1954
7765-9841	1954
9842-10565	1956
10566-11750	1957
11751-12512	1958
12513-12811	1959

SBW-6	
501-651	1955
652-690	1956
691-753	1957
754-780	1958

SBWD-6	
501-1453	1954
1454-3168	1955
3169-3903	1956
3904-4928	1957
4929-6039	1958

B-250	
501-3935	1956
3936-16026	1957
16027-23467	1958
23468-26897	1959
26898-29063	1960
29064-30260	1961

B-250 IND	
501-549	1958
550-557	1959
558-565	1960

B-275	
501-632	1958
633-12160	1959
12161-29467	1960
29468-39539	1961
39540-43119	1962
43120-46417	1963
46418-48976	1964
48977-51039	1965
51040-52932	1966
52933-54711	1967
54712-55546	

B275- PETROL	
501-1006	1961
1007-1110	1962
1111-1314	1963

B-275	
1315-1559	1964
1560-1747	1965
1748-1872	1966
1873-2043	1967
2043-2133	1968

B-275 IND	
501-517	1959
518-560	1960
561-590	1961
591-612	1962
613-637	1963
638-669	1964
670-702	1965
703-735	1966
736-749	1967
750-791	1968

B-414	
501-2547	1961
2575-14837	1962
14838-29646	1973
29647-42134	1964
42135-51663	1965
51664-53258	1966

B-414 PETROL	
501-2158	1962
2159-3994	1963
3995-5698	1964
5699-6344	1965
6345-6525	1966

B-414 NARROW TREAD	
501-661	1965
662-696	1966

434	
501-8592	1966
8593-16434	1967
16435-22079	1968
22080-286886	1969
26887-31617	1970
31618-31697	1971

434 PETROL	
501-1049	1966
1050-1962	1967
1963-2440	1968
2441-2951	1969
2952-3359	1970

434 HIGHWAY	
501-525	1968
526-543	1969
544-586	1970

434 NARROW TREAD	
501-666	1966
667-807	1967
808-960	1968
951-1169	1969
1170-1348	1970

B-450 International	
501-980	1958
981-3281	1959
3282-6201	1960
6202-9306	1961
9307-12427	1962
12428-15734	1963
15735-18324	1964
18325-21279	1965
21280-23794	1966
23795-25487	1967
25488-26845	1968
36846-27952	1969
27953-28791	1970

B-450 FARMALL	
501-656	1959
657-1979	1960
1980-3046	1961
3047-5637	1962
5638-6918	1963
6919-8041	1964
8042-9695	1965
9696-10373	1966
10380-10763	1967
10764-11026	1968
11027-11161	1969
11162-11512	1970

B-614	
501-507	1963
508-980	1964
981-2190	1965
2191-3016	1966
3017-3963	1967
3964-4448	1968

BTD-5	
501-510	1963
511-538	1964
539-563	1965
564-566	1966
567-667	1967

BTD-6 (40 HP)	
501-574	1953
575-1941	1954
1942-2350	1955

BTD-6 (50 HP)	
2401-3461	1955
3462-4640	1956
4641-6462	1957
6463-7931	1958
7932-9943	1959
9944-12037	1960
12038-14491	1961
14492-16124	1962
16125-18075	1963
18076-19372	1964
19373-19706	1965
19707-19923	1966